Great Whites of False Bay

South Africa

ISBN: 978-1-920094-52-2

A project of **New Voices Publishing**
www.newvoices.co.za

Great White with head above water

"We still do not know one thousandth of one percent
of what nature has revealed to us"

Albert Einstein

Acknowledgements

To the crew of the Blue Pointer, Rob, Sam, Jurgen and Wellington, thank you for imparting so much information and knowledge about the Great Whites and Seal Island, as well as the crew of the White Pointer, Chris, Poena and Jenna. To M. Zoghzoghi and Lee Nuttall, thank you for sharing your fantastic images. To my darling wife Jenny, thanks for your patience while I was out at sea – I love you.

The Great White Shark
(Carcharodon carcharias)

(image contributed by M. Zoghzoghi)

Introduction

Great White's of False Bay
South Africa

The Great White shark is undoubtedly one of the most well-known and most awesome apex predators of the seas and is also the largest predating fish on earth.
The Great White sharks visiting Seal Island annually display a unique behavior during hunting and predation of seals, by occasionally launching themselves out of the water in pursuit of their prey.
These breaches are spectacular and can be seen nowhere else in the world on such a consistent basis as in False Bay - South Africa, making it the prime destination for viewing these sharks in their natural environment.

This book illustrates some of these Great White sharks in action, and also provides the reader with some background information on False Bay, Seal Island, the Cape Fur Seal and its sharks.

I hope you enjoy the photography as much as I have enjoyed spending time with these sharks and capturing fleeting moments of their lives.

Dirk Schmidt
Cape Town, RSA

Sunset lighting the distant Hotentots Holland and Koeberg mountains overlooking False Bay

False Bay - South Africa

False Bay is located in the Western Cape Province of South Africa and can be described as a body of water which is defined on the western side of the bay by the finger-like projection of land, known as the Cape Peninsula.

It is a scenic bay, approximately 40 km across with the infamous Cape of Storms or Cape Point on its western shores, and Cape Hanglip on its eastern point.

False Bay is still, to this day, used as a safe anchorage during the winter storms. The name 'False Bay' was applied about three hundred years ago, by sailors who confused the bay with Table Bay to the north.

False Bay is also home to the South African Navy which is situated in Simonstown on the western side of the bay.

The eastern and western shores of the bay are rocky and mountainous; in places large cliffs plunge into deep water. The northern shore, however, is defined by a long and curving, sandy beach, reaching from the coastal towns of Muizenberg in the west and to the Strand in the east. Seal Island is the only island in the bay, and forms one of the main breeding sites for the Cape Fur Seal.

False Bay viewed from Boyes Drive above the town of Muizenberg and its popular surfing beaches

Dawn over Seal Island

Seal Island

Seal Island

Seal Island is a small, elongated, rocky land mass about 2 hectare in size with its long axis orientated roughly north-south, measuring approximately 400 by 50 meters. The island is located about 5.7 kilometers off the northern shore of False Bay, and is so named because of the number of Cape Fur Seals that occupy it. The island rises a mere six meters above the high tide mark and is off-limits to tourists, falling under the jurisdiction of the Western Cape Nature Conservation.

Seal Island is barren and devoid of any notable surface vegetation or beaches. The only noticeable evidence of humans ever reaching this desolate island is a solar powered light, remnants of a gas-powered lighthouse, as well as some rock inscriptions dating back to around 1930.

Human interaction on the island was limited to guano collection and seal harvesting. Guano collection came to an abrupt end in 1949, followed by a collapse in the fur seal product market in the 1980's, bringing to an end the practice of seal harvesting on the Island.

This has given the resident seal population time to recover to an estimated 60.000 individuals and the island now supports the largest Cape Fur Seal *(Arctocephalus pusillus)* colony in the Western Cape.

The growth of in the number of seals in and around the island has attracted an increasing number of Great White Sharks, making Seal Island one of the premier Great White locations in the world.

According to Rob Lawrence and Chris Fallows, Great White's become more evident at Seal Island at early- to mid-April each year, remaining in the immediate area until at least September. During this time, they predate heavily on Cape Fur Seals, with peak predation observed during June, July and August.

In the warmer summer months of December and January, Great White's are conspicuously absent from the waters around Seal Island, being sighted

closer to shore, cruising the backline of Muizenberg beach, Fish Hoek beach and the more easterly Macassar Beach.

Besides seals, a small population of African Penguins can also be found on Seal Island. Even though the seal population has increased substantially, the penguin population has remained stable with approximately 80 resident breeding pairs.

Other bird species which breed regularly on the island are White-breasted Cormorants, Cape and Bank Cormorants.

THE CAPE FUR SEAL
(Arctocephalus pusillus)

Cape fur seals can weigh up to 350kg and are considered the largest of all fur seals. The males have a rough mane, and powerful necks, and are much larger than the females, which only attain a weight of up to 90kg.

Mature bulls come ashore Seal Island in late October to establish territories which they actively defend, with older bulls often attempting to reclaim the same breeding territories they held the previous year, competing violently to establish a breeding harem of about 20 adult females.

The Cape Fur Seal has a gestation period of about 11.5 months (including a 3 to 4 months delayed embryonic development). Pups conceived the previous year are born on the Island between mid-November and mid-December with the bulls mating with their cows only 6 days after they having given birth. Seal pups are born black and molt for the first time at about 4 months of age, and suckle for almost a full year, taking solid food at an age of about 6 months.

It is estimated that on Seal Island about 2,000 Cape Fur Seals pups are born each season.
However, pup mortality is high within

Cape Fur Seals staying close to the relative safety of the island

the first year with a large percentage drowning, falling victim to infection, or being crushed by adult males. As a result, for several weeks each summer, hundreds of dead seal pups are found floating in the waters around Seal Island or are washed up on the nearby mainland beaches.

Cape Fur Seals leave Seal Island and travel far out to sea in search of food, having been observed as much as 50 kilometers offshore. Adult males can stay out at sea for months at a time, whereas adult females tend to remain at sea for a few days returning to suckle their pups.

Although seals, resident at the island, are generally not migratory, eight-month-old pups tagged at Seal Island have been recaptured at Cape Cross in Namibia, a staggering distance of over 1,600 kilometers. Cape Fur Seals hunt singly or gather on shoals of schooling fish and squids. In near-shore waters, they also may feed on small sharks, octopus, rock lobster, crabs, and other crustaceans. Fast and agile predators, the Cape Fur

Seals are capable of reaching speeds of approximately 15 km/hour, and dive to depths of about 200 meters, holding their breath for about 7-8 minutes.

Cape Fur Seals at the Island spend much of their daylight hours either resting on the rocky outcrops, aggregating in 'rafts' or cavorting in water shallower than 2 meters.

Aggregations of rafting and cavorting seals are seen primarily near the edge of the drop-off at either the south end of Seal Island or over the broad shallow bank at the north eastern side of the island. Individual Cape Fur Seals at Seal Island are especially vigilant of potential threats beneath the surface.

Collectively, these seals provide a constant sub-surface vigilance, making it very difficult for a shark or any other predator to approach (Lawrence, 2007).

Seals do sometimes survive Great White encounters, and often bear the scars to prove it. The picture above shows an adult female seal with fresh bite wounds emanating from a Great White Shark - note the circular jaw imprint.

16

Seal Islands viewed from the south. Note the group of seals on the right, congregating in a shallow area called the launch pad where they form groups before heading away from the Island.

A patrolling Great White near the launch pad

Rough seas near the area called the "Launch-pad" where seals gather in groups before leaving the island
(image contributed M. Zoghzoghi)

A "raft" of seals congregating in shallow water around Seal Island

A group of African Penguins forming part of the estimated 80 breeding pairs on the island

The northern end of Seal Island, with seals basking in the sun

A full breach, with the shark twisting side-ways

Seal Island

The Blue Pointer - African Shark Eco Charters

The triangle of fear

Small Great White approaching the boat

(images above and right by M. Zoghzoghi)

A full breach sequence takes about a second

The Great White shark *(Carcharodon carcharias)*

The Great White shark, also known as white pointer, white shark, white death and blue pointer, is an exceptionally large shark found in most coastal surface waters in all major Oceans. Reaching lengths of more than 5.8m (20 ft) and weighing up to 2,000 kg (4.200 lb), the great white shark is the world's largest known predatory fish.

It is beautifully streamlined to slip through the water with minimum effort. Its enormous size, powerful jaws, rows of large triangular teeth, and jet-black eyes make it one of the most feared creatures on the planet.
It is the ultimate macro predator.

Great Whites will hunt seals, porpoises, fish, and other sharks and will strike from any angle. It's large, powerful cutting teeth enables this shark to eat almost any living animal that ventures into its realm.

The Great White has been responsible for more attacks on humans than any other shark species.

According to the international shark attack file, worldwide records show that since 1580 more than 254 attacks on humans have been attributed to the Great White, of which 69 (or 27%) have been fatal. Surfers, divers, swimmers and kayakers being the most likely victims.

In False Bay, Great Whites are seen to breach right out of the water in pursuit of prey, combining speed, power and agility in launching themselves out of the water.

In most breaches, the sharks will protect their eyes from the prey's flailing claws and teeth by rolling them back into their sockets, giving the eye a whitish appearance as the nictitating membrane covers the eye.

White Sharks breaching
(images left and above contributed by M. Zoghzoghi)

Copepods seen on this shark's dorsal fin, are external parasites

Early morning predation, with seagulls in attendance. This is a sizeable shark

Natural predatory breach, note the seal underneath the shark

Shark appearance

Great Whites have a large, conical-shaped head. It has almost the same size upper and lower lobes on the tail fin, and commonly display counter-shading, having a white underside and a grey, sometimes brown dorsal area, that gives the shark an overall "mottled" appearance.

This type of coloration, makes it difficult for prey to spot the shark as it very effectively breaks up the shark's outline when seen from above, as the darker top half of the shark blends in with the darker sea.
In low visibility of 3-5m, even from an elevated vantage point of the boats viewing deck, it is difficult to spot Great Whites approaching even at a depth of only 2 meters. They simply blend in perfectly.

Early morning predation off Seal Island

Great White predation

The Great white sharks' reputation as ferocious predators is well-earned, yet they are not as was once believed, indiscriminate "eating machines".

Around Seal Island, they typically hunt using an "ambush" technique, taking their prey by surprise from below. Rob Lawrence and Chris Fallows have observed that the shark attacks on seals returning to the island most often occur in the morning, within the first 2 hours after sunrise. The reason for this is possibly that the shark's natural camouflage is optimized by the low light, and it is hard to see a shark close to the bottom at this time.

The success rate of attacks is about 50% in the first 2 hours, dropping to 40% in late morning and after which the sharks appear to stop hunting.

The Great White has a unique ability to breach out of the water.

This behavior can be seen nowhere else on earth on such a regular basis as in False Bay, and especially around Seal Island. This unique hunting behavior is mainly observed when hunting seals.

The shark rises from the depth of the sea, and like a missile, propels itself out of the water, attempting to grasp the seal. Usually the seal is completely unaware of the attack, Should the shark however miss the seal, the chase is on.

The unfolding scenes are an epic struggle of life and death, with the seal attempting to avoid the sharks jaws through agile jumps and direction changes, while the shark displays energetic persistence in its pursuit.

Whilst not all attacks are successful, surviving seals may be left wounded.

The series of pictures illustrated on the right show a young seal escaping the shark's jaws, but has already picked up some deep gashes on its left side, probably during the initial underwater chase.

The seal escaped the sharks jaws this time but already shows a few deep gashes inflicted by an earlier encounter with probably the same shark

A blood stain and a splash is often all that is visible from a sub-surface predatory event

Signature of a kill - seabirds marking the area

Natural predatory events are usually first detected at the surface by one or more of the following indicators
(Rob Lawrence - African Shark Eco Charters, Chris Fallows - Apex Shark Expeditions):

1.   **Incoming seals abruptly change course ad direction,**

2.   **Seals suddenly switch from porpoising to rapid zig-zag leaping,**

3.   **A group of traveling seals suddenly explode from the water in multiple directions in a star like pattern,**

4.   **A Great White Shark breaching with or without a seal in its mouth,**

5.   **A bloody splash, often accompanied by a spreading oily slick, or**

6.   **Seagulls or other seabirds wheeling over or plunging repeatedly toward a specific area of the sea.**

Attack in front of Seal Island - above image contributed by Lee Nutall

A surface lunge

A successful kill

Surface lunge with extended jaws, note seal lower left

Early morning predation. This shark attacked the seal below the surface, the large volume of blood in the water indicates that this was a successful predatory event

Note the mark on the lower dorsal fin of this shark, which could have been made by an adult seal or by external parasites (images above and right M. Zoghzoghi)

Predatory behavior of the Great White Sharks at Seal Island
(Adapted from Martin, A et al. 2005, and Lawrence R. 2007)

A typical **Polaris Attack** (shown on the left) is observed where the attacking shark performs a fast, vertical strike from beneath; often leaping partially or completely out of the water, with or without a seal in its mouth. If the shark managed to grasp the seal, it will typically shake its head violently from side-to-side, possibly stunning or causing the death of its prey. These types of attacks are extremely spectacular to observe.

After the initial strike where the shark failed to grasp the seal, the attacking shark may perform a shallow horizontal breach, generally re-entering the water with its prey in its jaws and forcing its prey underwater. The shark, having grasped the seal, shakes its head violently from side-to-side (laterally), possibly facilitating death or further severe injury. The lateral head-shake often proves fatal and is sometimes followed by a sidewards roll, causing one of the shark's pectoral fins to protrude from the water.

Great Whites can be seen performing a lateral bite or snap when a seal is 'working a shark' (a seals anti-predatory zigzag surface maneuvers) in an attempt to stay away from a pursuing shark, the attacking shark performs a sudden, swift lateral snap of its head, in an attempt to grasp the seal. Seals caught this way are generally grasped around the middle of the body, with the shark generally following up with a violent lateral head shake.

The lateral head shake maximizes the cutting efficiency of the shark's serrated teeth and results in the prey's rib cage being literally sawn apart.

The attacking shark may also inflict a killing bite on an incapacitated or injured seal, by performing a powerful bite to the seal's head or neck, instantly killing its prey. A killing bite may be followed by a surface grasp, in which the shark, having killed or incapacitated its prey, performs a seemingly slow grasp of the floating seal, sometimes raising its head partially out of the water to grab the seal. This is usually followed by a less violent head shake or by dragging its prey along underwater before the shark consumes its prey.

A dead seal will inherently float at the surface and it is where the shark will calmly and efficiently feed upon it. This is typically performed in two or three massive bites. The shark is usually seen exhibiting a number of lateral head-shake's to saw apart the carcass. Surface feeding generally lasts less than a minute or two.

The seal's entrails are rarely eaten by a Great White, providing food for seagulls and other seabirds which compete vigorously for any leftovers. Seagulls diving or settling on water near a seal colony are often the first sign of a Great White's kill and are even seen before a kill in anticipation of a meal.

Breaches on a calm day, off Seal Island
(image left and above contributed by M. Zoghzoghi)

Great White patrolling near the boat, the size of the fin shown above the surface often does not reflect the size of the shark.

A high breach, meters from the boat White Pointer II,
of Apex Shark Expeditions

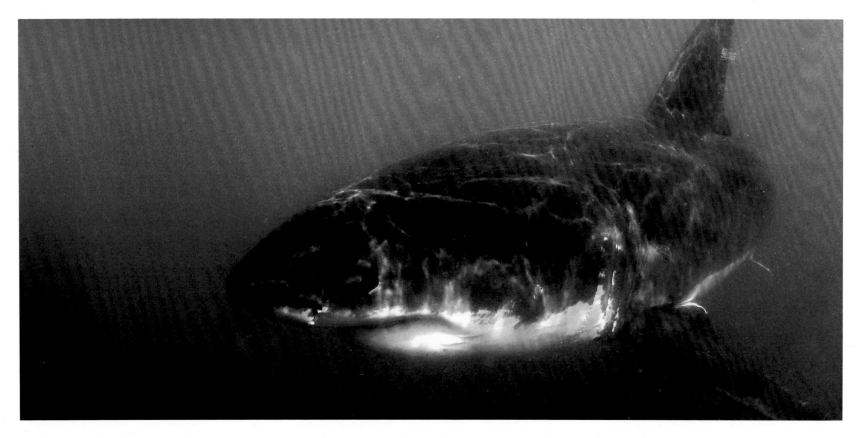

A Shark called "Cruella"

A 4.5m female Great White arrived at the Island in late April 2008 which displayed an extraordinary level of aggression and savvy when near the boat. To such an extent, that besides earning her nickname, she often surprised the boat staff handling the bait.

The purpose of the bait is not to feed the shark but to attract and keep the shark interested in the or near the boat, therefore the bait is pulled away from the shark towards the boat. Baits usually consist of tuna or large fish heads caught during commercial fishing in the off-season.

During the 2008 season, Cruella earned her reputation by taking everything in the water, and even exhibiting a breach near the boat, causing all dive and bait operations to be suspended until she lost interest and moved on. One memorable trip, I remember she took 2 baits without the boat staff even having a chance. Coming up directly beneath the bait, or sneaking outwards from underneath the boat she got her prize, as though she calculatingly worked out how to get to the bait without it being pulled away from her.

Towards the end of the 2008 season, she calmed down and disappeared again as suddenly as she arrived.

*A massive female Great White
near the bait in the late afternoon*

Shark Awareness

Considering the number of people who use the ocean on a daily basis, the actual number of attacks by Great White Sharks is extremely low. People are not the natural prey of the Great White Sharks. Cape Town has experienced 6 attacks in the last 5 years, given that hundreds of surfers use the sea on a daily basis, and knowing that Great Whites frequent their favorite surf spots, the chance of being attacked by a Great White stays relatively remote.

In fact, considering the number of regular surfers in our Cape waters, and given the extraordinary eyesight of these sharks, it is entirely possible that for every 100 surfers at least 1 is "checked" out by a passing shark without the surfer having noticed.

Everyone entering the ocean is reminded that they do so at their own risk and discretion, knowing that they are entering a wild environment where sharks naturally live, and have lived for millions of years.

We choose to enter their territory; they are not coming into ours – please respect that they were there before us, and it is their domain.

(images left and above contributed by M. Zoghzoghi)

A sizable female Great White cruising past

(image by M. Zoghzoghi)

(above image contributed by M. Zoghzoghi)

Great White approaching the bait

A large Great White on the chase

A spectacular breach off the stern of the White Pointer II
(image by M. Zoghzoghi)

Shark tourism and cage diving

Shark cage-diving involves a steel cage being lowered into the water, either next to the boat or on a tether, for group of tourists or individuals wishing to experience and learn about Great Whites up close. These individuals brave the cold water and have the opportunity to experience the most awesome and majestic predators

alive. They learn about these creatures, often changing their mindset about Great Whites (simply being mindless eating machines), to respecting this apex predator in its natural environment.

Having said this, there is no other method of safely viewing Great Whites underwater, considering the risks of free or scuba diving. Great Whites cannot be

kept in captivity successfully. Without Great White tourism and shark tour operators, many of us would not be able to ever view and experience Great Whites. They would simply remain surrounded in myth and irrational fear, with movies like JAWS being our main reference.

I believe that cage diving has helped save Great Whites in debunking many

The White Pointer II, Apex Shark Expeditions

of the misunderstandings regarding food – a potentially dangerous situation. It is claimed that certain methods of attracting sharks to the boat or cage, such as when bait attached to a rope is pulled towards the divers in the cage, drawing the shark in to the divers proximity, exacerbates this problem.

If this was so, and Great Whites become accustomed to people (clad in their black neoprene wetsuits), we would have seen a massive increase in shark attacks in False Bay, where a major seal colony is situated not even 6000 meters from Muizenberg beach, a favorite and heavily frequented surf spot, is situated.

A 6000 meter swim for a hungry Great White is nothing considering they may possibly swim or even migrate for thousands of kilometers.

Furthermore, it has been shown in numerous studies, that humans are not on the Great Whites preferred menu. Surely it would stand to reason, had the shark attacked a human with purpose, the human would be consumed? Images contained in this book show Great White with purpose, attacking to kill and feed. Should a human be attacked in this manner, it would be unlikely that we

Trying to keep a White Shark's interest near the boat is hard work

would survive the attack.

After all, would a Great White not smash the puny, little cage containing the human morsel if he really associated humans as food?

Even if the shark's feeding stimulus has been aroused and it actively swims around the boat, why does it not attack the cage? Conditioning sharks to humans, in my opinion, does not increase the likelihood of attacks or human predation. Unless it can be scientifically proven that attracting sharks through repetitive human interaction, a cognitive behavior is being developed by Great Whites to associate humans with food, and thereby increase attacks on humans, or change the attack modus to feeding versus biting – cage diving and shark tour operations should be allowed, albeit in a regulated fashion which governs marine safety and ecological considerations.

Large Female Great White

Large Great White breaking surface

Only a limited number of permits have been issued for 3 operators, two operating from Simons Town and the other from Gordons Bay. These permits allow the operator to interact with the Great Whites near Seal Island.

(image by M. Zoghzoghi)

The seal decoy after being "hit" by Great Whites. Whilst the decoy shows bite marks, the sharks usually instantly recognizes it as non-edible and discards it.

The seal decoy, consisting of some foam and carpet is towed for a maximum of 40 minutes as not to interfere with the sharks natural hunting patterns and is towed using a thin nylon line.

(Left) Wellington Baba, boat assistant on The Blue Pointer, re-attaches a thin line after the seal decoy was taken by a Great White, breaking the line during a broach. The seal decoy has been shaped in the form of a seal.

(Above) the crew attaching the cage to the side of the boat and getting ready for divers.

(Left) The crew of the Blue Pointer, Jurgen Batsleer and Sam Croome preparing the cage for the days dive session. Jurgen is a qualified marine biologist, and Sam is a Class III navigator

(Below) A diver in the cage attached to the boat

Jaws agape, the bait is again pulled away from the oncoming shark

A Great White approaching the cage

Large White shark breaching near Seal Island

The need for conservation

More than 100 million sharks are harvested every year, with many species on the brink of extinction or population collapse. High sea fishing allows for the virtual uncontrolled hunting of sharks, which are processed for their fins, or end up in shredded fish products such as fish patties, pet food etc.

These magnificent creatures, which roam the oceans, cannot defend themselves against scrupulous humans, with their nets, long lines and steel hooks. The only way to save sharks, in my opinion, is to generate an economic need for live sharks – tourism! Fortunately, shark tourism is becoming a thriving industry world-wide, with many operators specializing in shark trips. A dead Great White, which maybe fetched a few hundred dollars, now is worth thousands as tourists, dedicated divers, scientist and environmentalists support sharks – as the more people experiencing these magnificent creatures, the better chance they have for survival since a demand is created by people wanting to see a live shark.

Survival however starts with everyone at home – do not use or support shark products, and preserve the species for our children and maintaining the natural biodiversity of this planet.

Simons Town harbor
(image by M. Zoghzoghi)

About the Author

Dirk Schmidt Ph. D

Dirk's passion for sharks started as a teenager when at 17 he compiled his own extensive research dossier on over different 100 shark species, their habits, feeding behavior, biology and distribution.

As a qualified dive master, he started free-diving with sharks, long before it became fashionable, realizing at an early age that sharks where not the mindless killers often portrayed and sensationalized in the press of the time.

His book, Great Whites of False Bay, is a culmination of his life long fascination of sharks, and in particular the enigmatic Great White Shark – Carcharodon Carcharias. He portrays this apex predator in an unbiased pictorial review, its predatory behavior and specialized hunting techniques, interaction with other members of its species as well as the sheer beauty and power of this magnificent creature.

His images of Great Whites allow us to share a rare insight into Great White Shark behavior, and witness an aspect of the natural cycle of life at sea.

In preparation for this book, Dirk has spent more than 300 hours in close proximity of Great White Sharks, studying, observing, learning and capturing images of this magnificent apex predator.

He lives with his wife Jenny, in Simons Town which is located on the far western edge of False Bay, less than 20 km's from Cape Point.

(Above) Seals rafting together in the relative safety of the shallow water near the island   *(right)* image contributed by M. Zoghzoghi

References:

1. Chris Fallows, Apex Shark Expeditions, 2008

2. Lawrence, R., African Eco Charters 2007. *Life history of the fur seal, South Africa,* www.ultimate-animals.com

3. Martin, R.A., N. Hammerschlag, R.S. Collier, and C. Fallows. 2005. *Predatory behaviour of White Sharks (Carcharodon carcharias) at Seal Island, South Africa.* Journal of the Marine Biological Association of the United Kingdom, 85: 1121-1135.

4. Martin, R. Aidan. 2003. *Biology of Sharks and Rays* World Wide Web Publication

African Shark Eco Charters and Apex Shark Expeditions

ASEC is owned by Rob and Karen Lawrence. Rob has been working with sharks on a daily basis since 1989 and has specialized in leading and operating shark expeditions and trips since 1996. He has a genuine love of for these predators and this is conveyed when you come along on his trips.

ASEC started out very small in 1996 and only ran a few trips a year, the reason for this was that they were learning all the time about our new, unique area as well as the fact that we did not want to open it up to exploitation. This has happened at other shark sites around the country.

For this reason they believe that Seal Island, False Bay is the best place on the planet to see white shark predatory behavior.

There are currently 3 operators in False Bay. One of these is Chris Fallows of Apex Shark Expeditions. We have worked along side Chris for years, and feel that they run their operation with the same ethos and vision as we do. Unfortunately there are plans to issue more permits by Marine and Coastal Management. We think that this would be extremely unwise. This area is incredibly sensitive. It takes years of working with sharks to understand predatory behavior etc. Handing out permits left and right in the aim of "transformation" would be detrimental to the sharks.

The industry is strictly controlled and monitored. They belong to the White Shark Protection Foundation. It acts as a body to protect great white sharks and monitor that operators are functioning according to their permits code and conduct, whilst looking out for the industry as a whole. Rob and Chris and their respective crews are extremely professional and safety conscious, and maintain the highest ethical standards set by their permit, allowing them to interact with Great White sharks.

Rob Lawrence

African Shark Eco-Charters can be contacted through www.ultimate-animals.com